LIFE SKILLS
IN THE PACIFIC

Basic Maintenance

Eron Hagunama

OXFORD

Contents

Introduction

Basic Maintenance is about carrying out maintenance and repairs on homes and buildings. It contains step-by-step procedures for carrying out repairs and maintenance on roofs, walls, floorings, and furniture. It also covers plumbing repairs such as leaking water taps, blocked sinks, toilets and broken pipes. Care and maintenance of handy tools and plumbing tools are emphasised as well.

This book is written to complement the newly introduced 'Making a Living' subject for upper primary students. It supports the philosophy of Education, which emphasises relevant education and self-reliance. It teaches students useful skills and relevant knowledge to enable them to become useful members of their local communities.

Strand: *Better Living*
Sub Strand: *Care and Management*
Outcomes: 7.2.2: *assess home and school buildings to identify areas that require maintenance, repair or other improvements and undertake appropriate action*
8.2.2: *work collaboratively with others to select and undertake a project based on identified needs within the school or community*

I encourage students not only to read this book but also to use the information in this book to develop useful skills that will assist them in their life.

Eron Hagunama

Common home maintenance problems

The chart below serves as a quick reference to the most common home maintenance and repair problems that occur either in the home or in local community buildings such as schools, community halls, churches and so on.

Beside each problem is listed:

- what may have caused the problem
- how to fix the problem
- a page reference to step-by-step instructions for carrying out the repairs/maintenance.

Repair or maintenance problem	Possible cause	Repair instructions	Page reference
Wet spots on the ceiling	Loose roof nails or rust holes	• Replace the roof nails • Patch holes with bituminous compound	**14**
Water leaking or overflowing from the roof gutter	Rust holes or a clog in the gutter or downpipe	• Patch the rust holes • Clear the clog	**17–18**
Water leaking from kunai or morota roofs	Kunai grass on roof rotting or attacked by insects	• Replace with new kunai grass or morota leaves	**19–20**
Broken fibro wall	Hard hit on the fibro wall	• Patch the broken section with a new piece of fibro	**21**
Dirty walls and ceilings	Dirty finger marks, spider webs, mould or dust	• Clean the walls and ceilings	**25**
Blind walls losing colours or rotting	Exposure to sun and rain	• Replace with new blind	**27**

Repair or maintenance problem	Possible cause	Repair instructions	Page reference
Broken window louvres	Hard hit on the louvres	• Remove broken pieces and replace with new louvres	**32**
Broken or damaged floor	Hard hit or termite attack	• Repair the damaged section	**34**
Dirty floors	Dust, mud, frequent use of the house	• Sweep or mop the floor	**35**
Door coming loose	Loose screws from door hinges	• Tighten the screws firmly	**39**
Broken doorsteps	Frequent use and exposure to sun and rain	• Remove and replace the damaged section of the step	**41**
Broken furniture	Frequent or careless use	• Mend the furniture	**46**
Leaking water taps	Worn out washer	• Replace with new washer	**48**
Blocked sink causing overflow in the kitchen	Blockage in the sink and pipes	• Clear out blockages in the sink and pipes	**50**
Blocked toilet, water in the toilet bowl rises rather than recedes	Block in the pipes	• Clear out the pipes	**52**
Water coming out from a broken pipe	Broken water pipe	• Repair the broken water pipe	**53**

Tools and equipment

A basic tool kit helps you to carry out your own repairs at home, at school and in your community following the 'do it yourself' steps suggested to you in this book. A basic maintenance tool kit includes tools such as a claw hammer, chisel, clamp, try square, crosscut saw, plane, tape measure, screwdrivers, hand drill, sanding block, and paintbrushes.

Tools cost a lot of money, but you do not need to buy them all at once. You can buy your tools one at a time, and you do not need to have all the tools listed below. However, it is essential to have basic tools. It is very important to look after the tools you have.

Here are some useful tools that will help you in carrying out your maintenance and repair works effectively.

Maintenance tool kit

Claw hammer

Used for striking nails into wood. The claw part of the head is used for pulling nails out of wood.

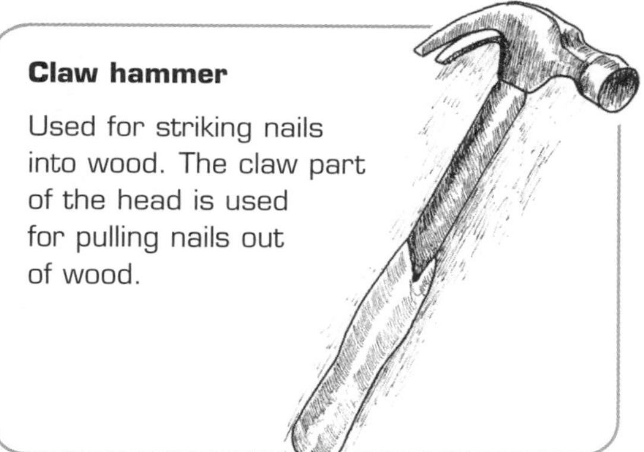

Screwdriver

Used for driving screws into wood or metal. There are two different types of screwdriver, flat head and Phillips head.

Saw

Used to cut wood and board. There are two different types of wood saw. A crosscut saw is for cutting across the timber grain, and a ripsaw is used for ripping or cutting along the timber grain.

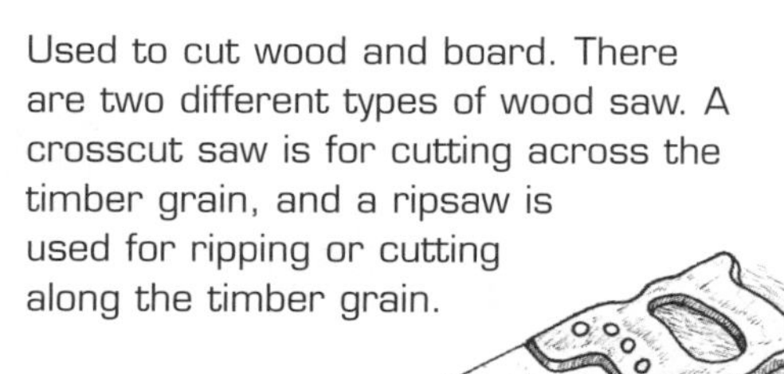

Plane

Used for smoothing the surface of timber and reducing the thickness of the timber to the required size.

Chisels

Used for reducing the thickness of wood where planes cannot.

Clamp

Used to hold wood or metal firmly while it is cut or drilled.

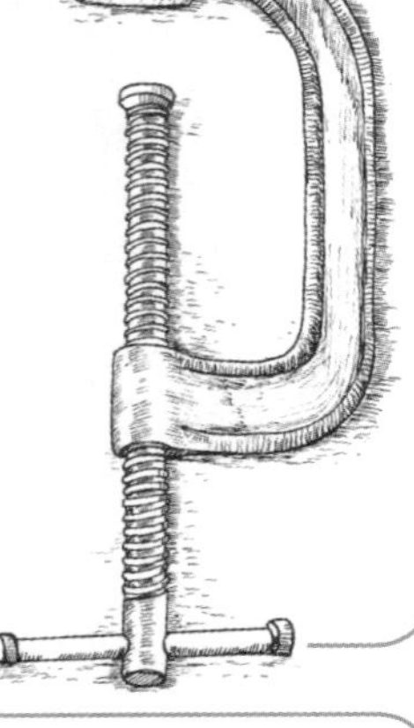

Wooden mallet

Used to drive a chisel into wood.

Drill and bit

Used to make holes in wood. The drilling piece at the end of the tool is called a 'bit'. There are different shapes and sizes of bits used for different purposes.

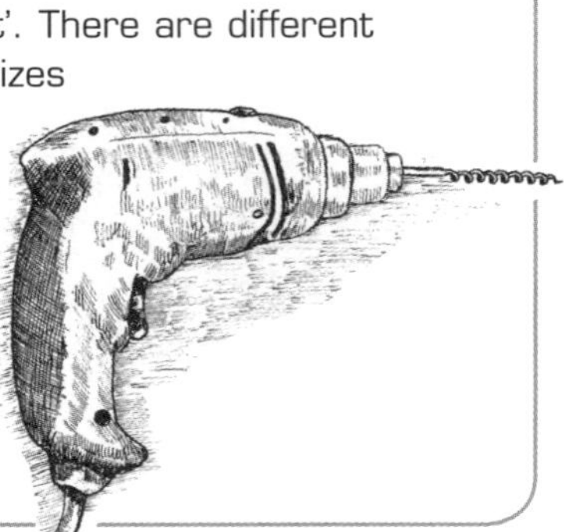

Bush knife

Used for cutting young trees and bamboos and clearing bushes.

Axe

Used for felling trees and splitting logs.

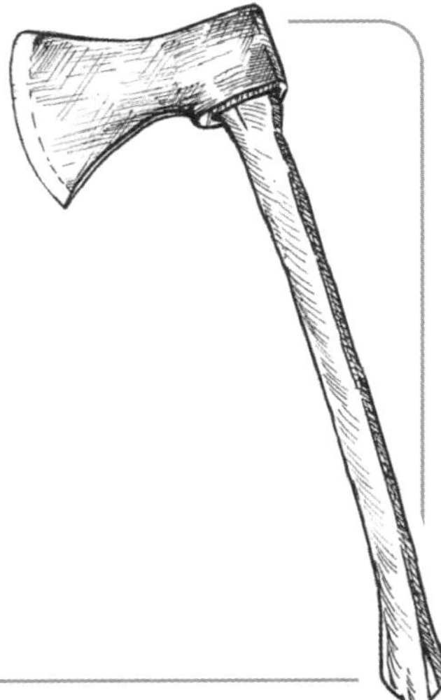

Paintbrush

Used for applying and spreading paints on wood and boards. It has a wooden handle and fine bristles to spread the paint. Always wash the brush in turpentine after painting.

Toolbox

Used to keep your tools in. It is easier to carry your tools to your work site if you keep them in a toolbox.

Bucket

Used for fetching and holding water.

Mop

Used for wiping a wet floor clean.

Broom

Used for sweeping dust and dirt from the floor.

Sanding block

A piece of wood with sandpaper wrapped around it, used for smoothing the surface of wood before painting or varnishing.

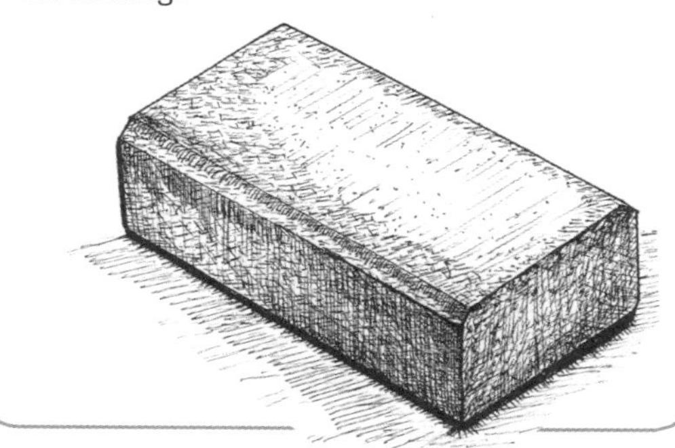

Hacksaw

Used for cutting metal pipes.

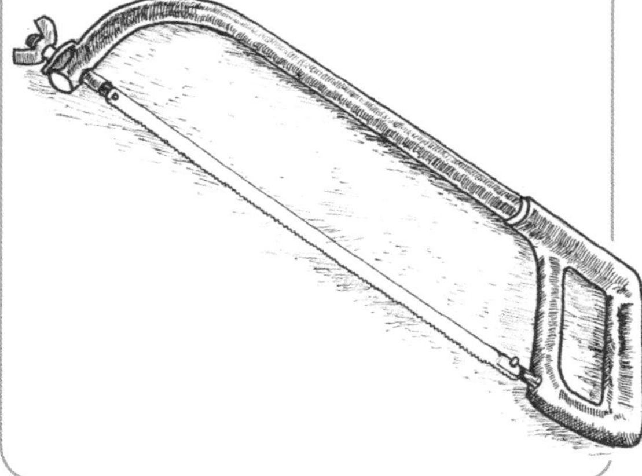

Tenon saw

Used in light jobs for cutting wood and thin boards.

Wire brush

Used for removing rust from metals.

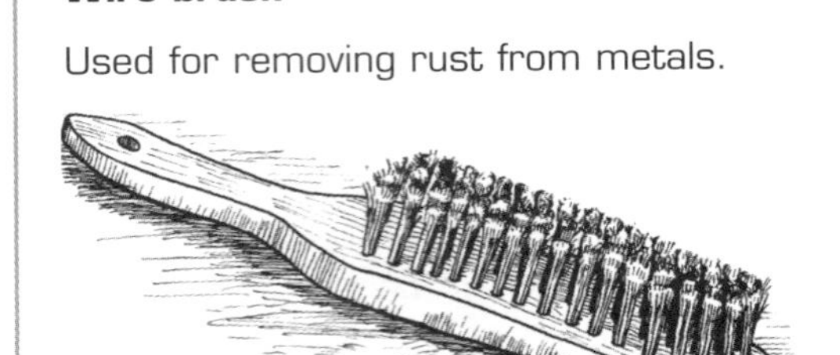

Tape measure

Used to measure materials.

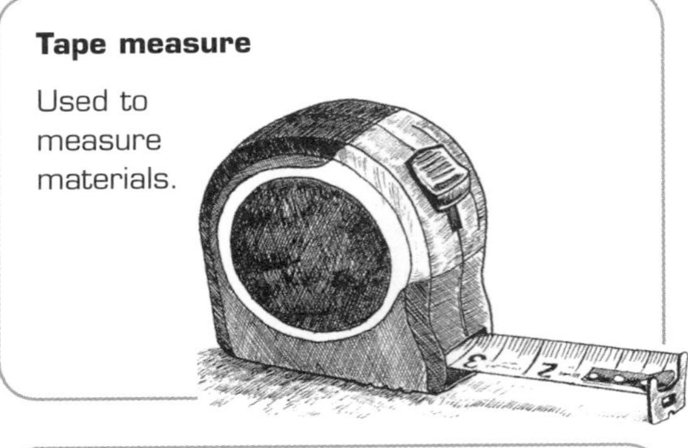

Toilet brush

Used for cleaning toilets.

Pliers

Used for holding and cutting wires.

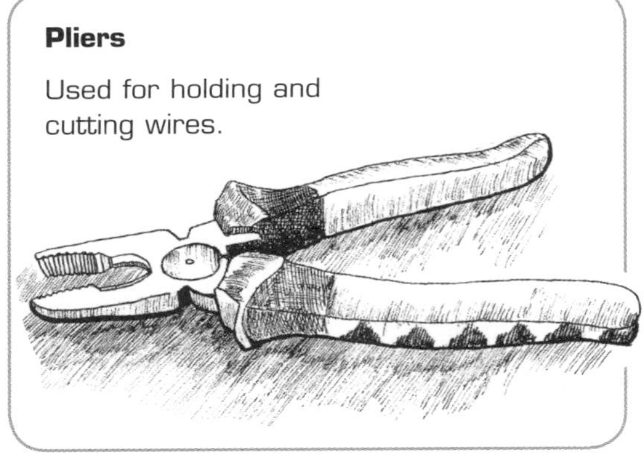

Scrubbing brush

A strong brush used for scrubbing concrete and unpolished wooden floors.

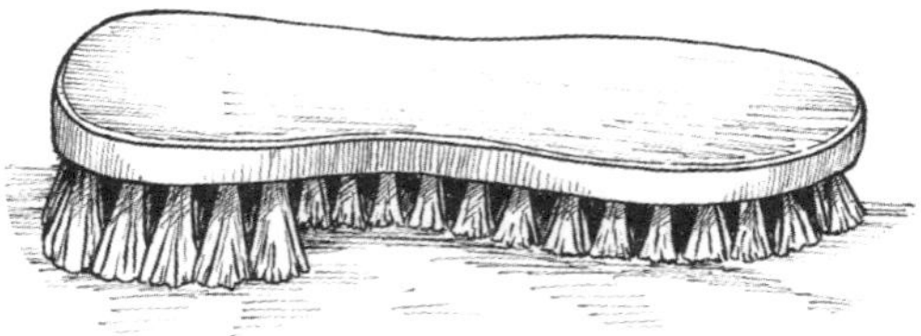

Shifting spanner

Used for loosening and tightening nuts on bolts.

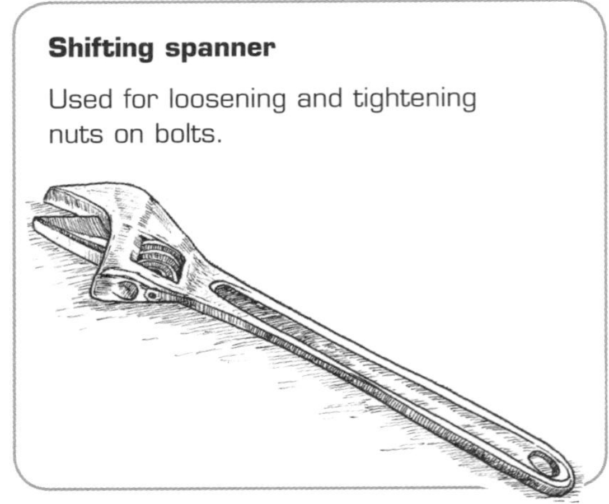

Nail punch

Used for driving the head of nails into wood.

Tinsnips

Used for cutting sheet metal.

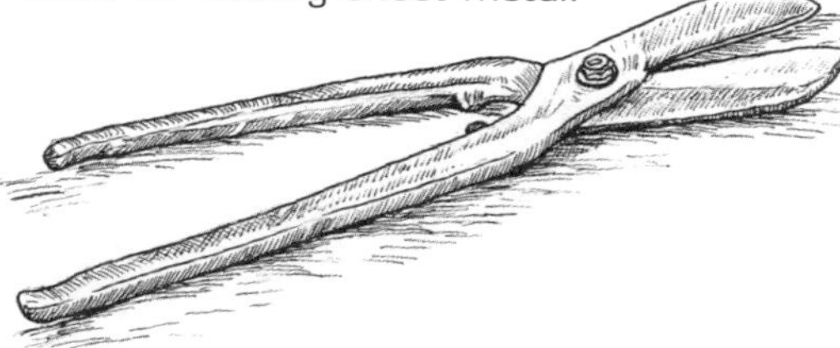

Spirit level

Used to test if a surface is flat and even.

Putty knife

Used for applying wood filler or tar into gaps or holes in wood or sheet metal.

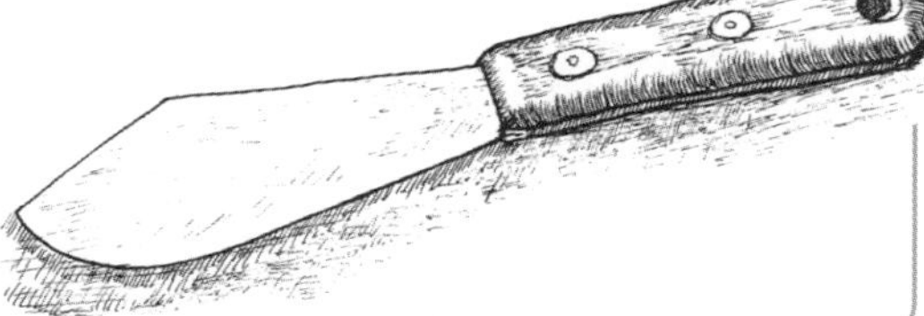

Glass cutter

Used for cutting window glass to the required size.

File

Used for filing wood or metal. There are two types, wood files and metal files.

Maintenance and repair of tools

Woodwork tools and equipment are expensive. Tools need to be cared for and stored well to keep them in good working condition so that they last as long as possible.

General hints on caring for tools

- ✓ Clean and oil the tools thoroughly after use.
- ✓ Store tools in a secure place like a toolbox or a cupboard.
- ✓ Only use tools for the purpose for which they were meant to be used.
- ✓ Repair the tools if they become damaged so that they can be used safely.
- ✓ Keep a stock book to record the number and condition of tools and equipment.
- ✓ Create an effective borrowing system to make sure tools are returned after use.

Plumbing tools and equipment

Force cup

Closet auger

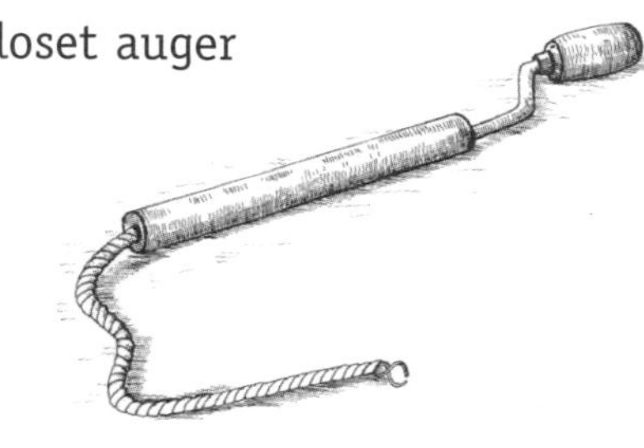

Pipe clamps

Sheet rubber

Drain cleaner

Electrician tape

Locking pliers (vice grips)

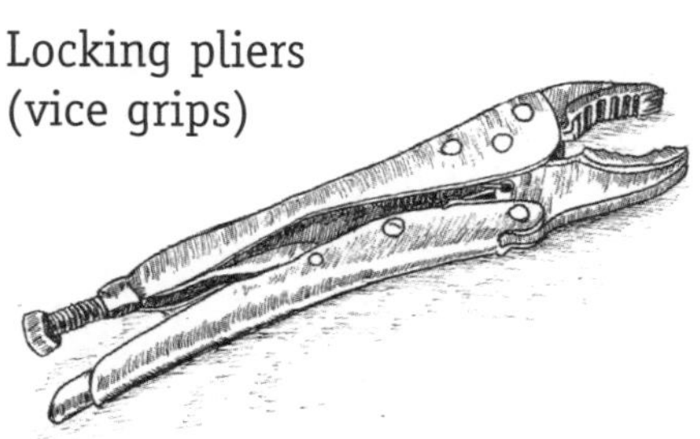

Drain auger

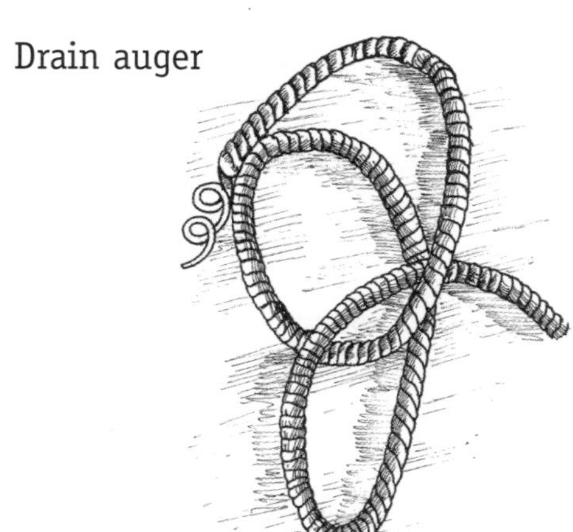

Adjustable wrench

Maintenance to roofs

Repairing a leaking roof

A wet spot on your ceiling when it rains is usually the first sign that there is a leak in the roof. Most roof leaks are caused by loose roof nails, rusty holes, or damage to the overlapping of corrugated roofs.

What you will need

- ✓ hammer
- ✓ roofing nails
- ✓ bituminous roofing compound (tar)

INSTRUCTIONS

Check where the water is coming in.

If the roofing nails have come loose, try a new, larger roofing nail.

3

Nail in new nails to secure the roof.

4

Patch the old nail holes with a thick coating of bituminous roofing compound (tar). Rust holes in corrugated roofs can be patched in this way too.

For leaks from roofing nail washers, replace with a new roofing nail or patch its underside with the bituminous compound.

Repairing an iron roof

What you will need

- ✓ wrecking bar
- ✓ tinsnips
- ✓ hammer or screwdriver

INSTRUCTIONS

1

Remove the old nails or screws from the old sheet of roof iron. Use a wrecking bar to remove the nails. Place a piece of wood under the bar so that you don't damage the iron.

2

Lift the bottom sheet at an angle to clear those on both sides and slide it downwards.

3

Turn the hollows of the corrugations at the end of the sheet so that they are positioned at the apex of the wood.

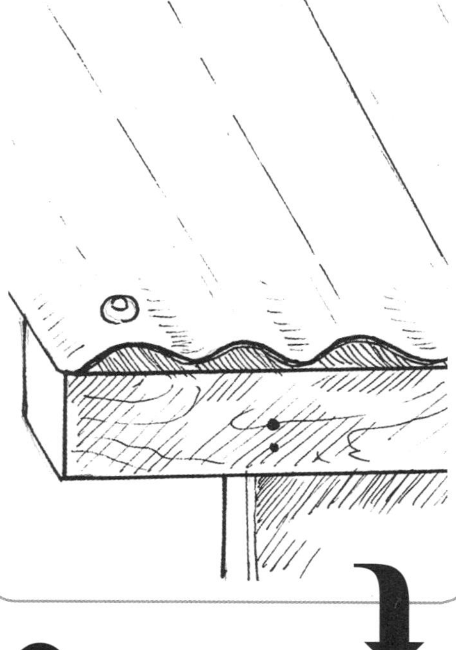

4

Cut a small triangle off the new sheet to stop it from binding as it slides into position.

Repairing an iron roof continued ...

5

Insert the lower of the new sheets under one old sheet and cover the other. Push it until it is level with the others.

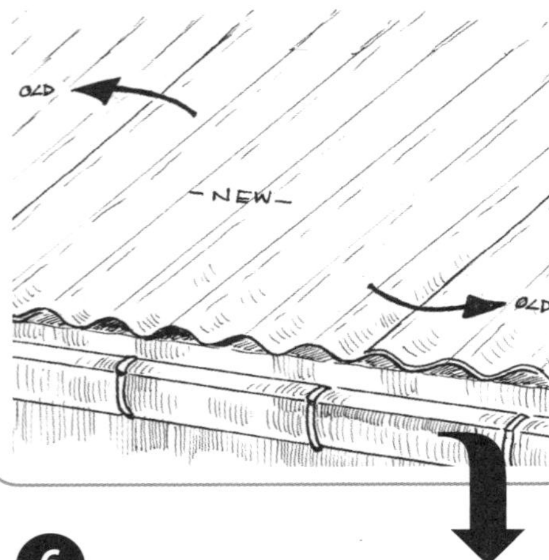

6

Punch holes in every second corrugation and fix nails or screws firmly.

7

Gently lift up the ridge cap to insert a new top sheet.

8

Insert the new sheet and push it up until the top end fits under the capping.

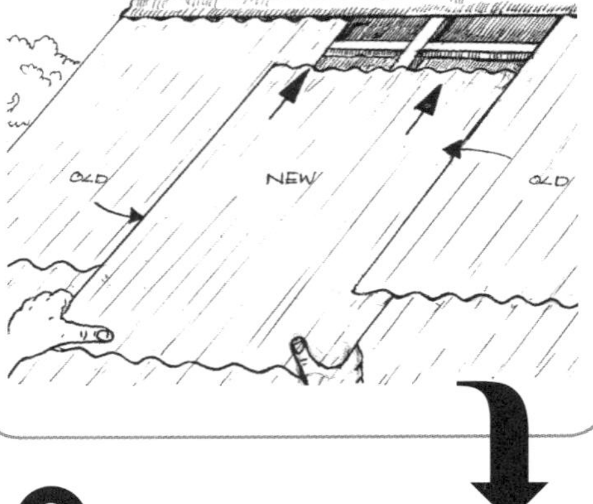

9

Move the sheets to make sure the corrugations fit. Nail every second crest at the ends and every third crest in the middle.

Repairing roof guttering

Proper care of guttering makes your gutters and downpipes last longer.

What you will need

- ✓ hacksaw
- ✓ hammer

INSTRUCTIONS

1

Clean out gutters every three months to remove dead leaves. This prevents the clogging of downpipes and stops water overflowing from gutters when it rains.

2

If you have bituminous paint, coat the inside of the galvanised guttering to protect the gutter from rusting.

3

Keep the outside of the gutter well painted.

4

If the gutter is damaged, remove it and replace it with new guttering. If you find that only one or two sections of the guttering have rusted through, cut away the damaged section with a hacksaw before loosening nails and brackets.

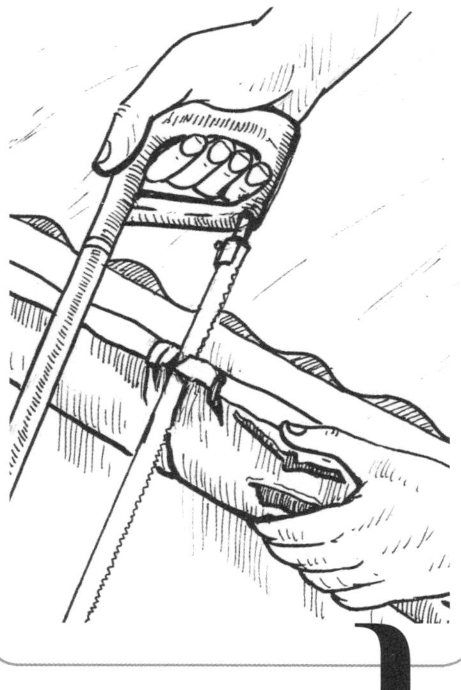

5

Nail the joined guttering firmly onto the fascia board.

Repairing downpipes

The downpipe should not get clogged if the gutter is cleaned regularly, but if there is a blockage, probe down the pipe with a long stick.

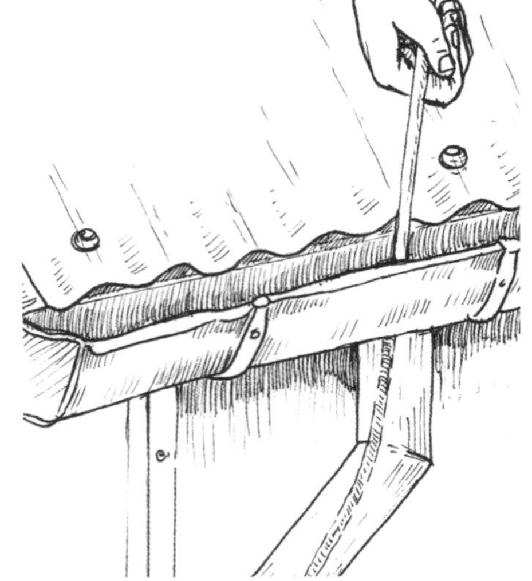

What you will need

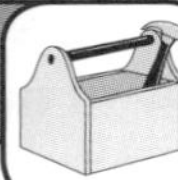

 mesh wire

- ✓ hacksaw
- ✓ tinsnips

INSTRUCTIONS

1

Cover the ends of the downpipe with mesh wire to prevent clogging.

2

To replace a broken section of downpipe, remove the whole pipe from the wall.

3

Cut the damaged section of the pipe with a hacksaw and tinsnips.

4

Cut along the seams of the new section of pipe and fit it around the outside of the top pipe and the inside of the bottom pipe.

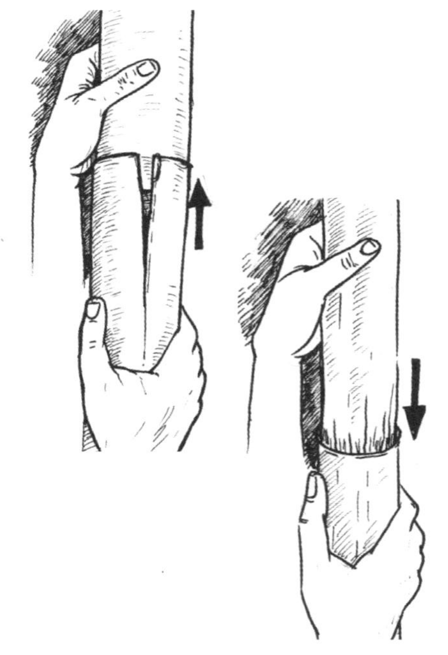

Changing kunai grass roofs

Kunai grass roofs are made from a bush material called kunai grass. These roofs are common in traditional houses in the Highlands and in some areas of the lowlands of Papua New Guinea.

A kunai grass roof generally lasts for up to 5–6 years. When leaks start to appear in the roof, the old rotting kunai grass must be replaced with new kunai grass.

What you will need

- cutting tool
- claw hammer

INSTRUCTIONS

1

Cut loose and untie the rope supports from the roof. Pull out any nails used to nail the supports.

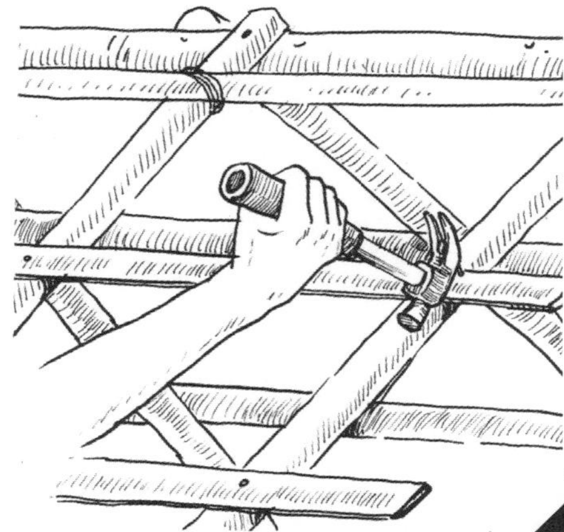

2

Remove the old kunai grass from the roof. Start from the top and work downwards.

3

Replace with new kunai grass and tie it firmly onto the roof rafters.

Changing morota roofs

Morota roofs are made from sago palm leaves that are either woven or stitched onto bamboo splits. These roofs are used on traditional houses in coastal and swampy lowland areas in Papua New Guinea. The roofs last up to 5–6 years. When they start to rot or show signs of insect damage they need to be replaced with new prepared morota leaves.

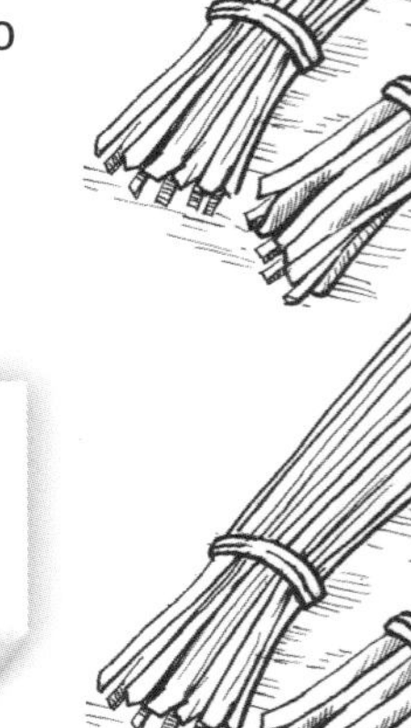

What you will need

- ✓ cutting tool
- ✓ claw hammer

INSTRUCTIONS

1 Cut loose and untie the ropes around the old morota leaves. If the supports were nailed, pull out the nails.

2 Remove the old morota leaves from the roof rafters. Start from the top and work downwards.

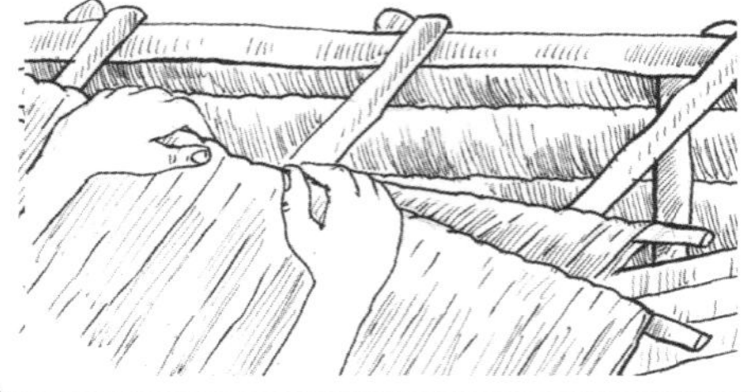

3 Replace with new prepared morota leaves and tie them firmly onto the roof rafters.

Note: *New morota must be prepared and left in the sun to dry before it is used. Morota can be woven or stitched onto bamboo splits.*

Maintenance to walls

Repairing broken fibro on external walls

The external (outside) walls of most permanent houses are made of fibro. Damaged fibro can become a big problem if it is not repaired, so if an area of fibro is damaged it is wise to repair it quickly before it becomes worse.

What you will need

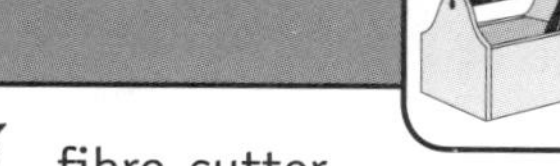

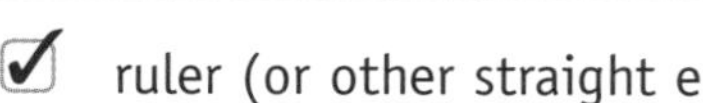

- ☑ ruler (or other straight edge)
- ☑ pencil
- ☑ crosscut saw
- ☑ tape measure
- ☑ fibro cutter
- ☑ hammer
- ☑ fibro nails
- ☑ drill with 2 mm drill bit

INSTRUCTIONS

1

Locate the studs and nogging around the broken pieces of fibro. The line of the nails shows the location of the studs.

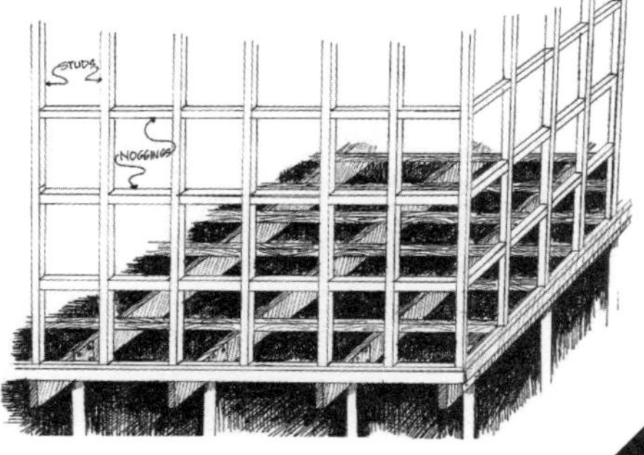

2

Using a straight edge (timber or ruler) and a pencil, mark out a line on the fibro which is in the middle of the studs and nogging.

3

Using a crosscut saw, scrape out a 'V' cut at least one half of the thickness of the fibro sheet. Use a piece of timber to guide the saw.

Repairing broken fibro on external walls continued ...

4

Pull the broken section of fibro off the wall.

5

On removing the piece of fibro, take its measurements and transfer these measurements to a new piece of fibro.

6

Cut the new fibro to the right size using a fibro cutter.

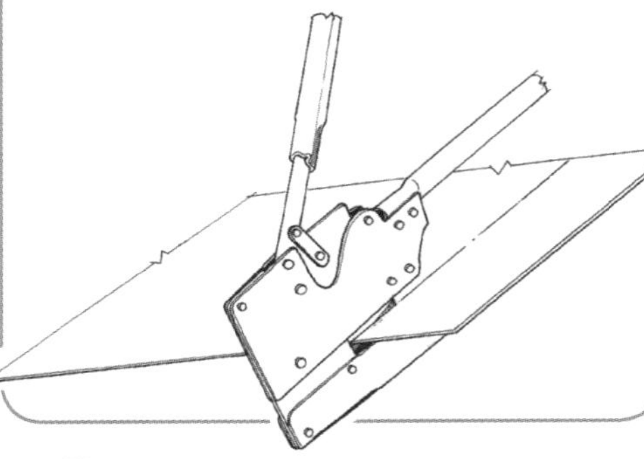

7

Place the new piece of fibro in position and nail with fibro nails.

8

If the section is required to be waterproof, a piece of galvanised iron cover strip could be put on the horizontal joint.

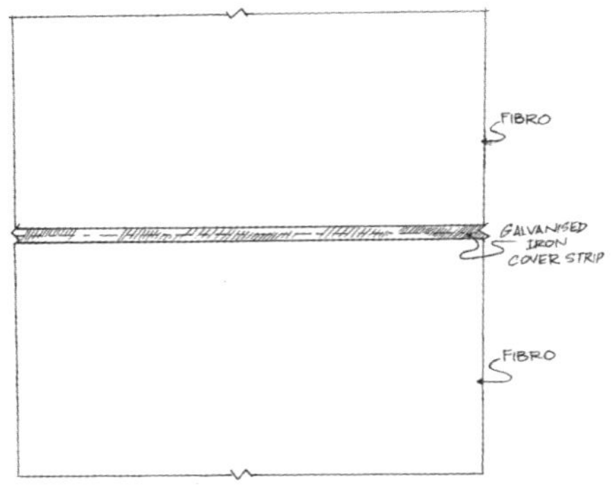

9

Finally, nail the horizontal cover strips into place. (Use a 2 mm drill bit to drill the pilot holes.)

Repairing masonite or plywood internal walls

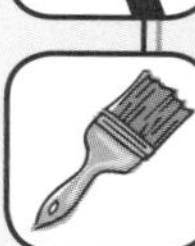

Most modern houses have two sets of walls: the external (outside) walls and the internal (inside) walls. The main reasons for having internal walls are to help keep the house cool, to make the house more secure and to make it look more attractive.

There are three main materials used for inside walls:

- masonite
- plywood
- fibro.

These materials can be bought in sheets from all local hardware stores.

What you will need

- ☑ centre punch
- ☑ pencil
- ☑ saw
- ☑ hammer
- ☑ putty
- ☑ paintbrush and paint

INSTRUCTIONS

1 Remove the cover strips from the broken walls.

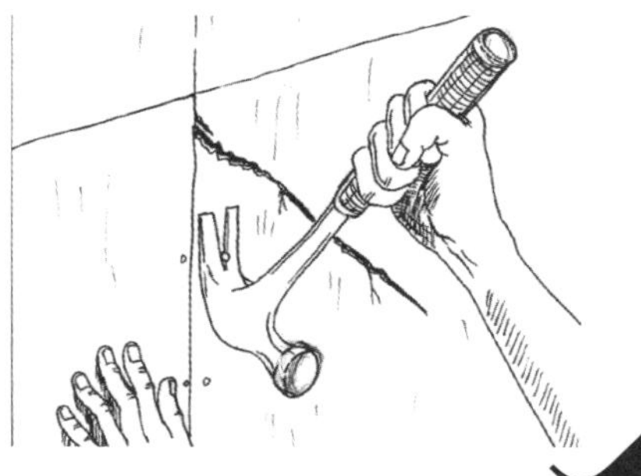

2 Punch the nails with a centre punch.

3 Carefully remove the old sheet of masonite or plywood.

Repairing masonite or plywood internal walls continued ...

4

Trace the shape and size of the old sheet onto the new sheet and cut it to the correct size.

5

Fit the new sheet onto the wall to check that it sits properly.

6

Nail the studs and nogging.

7

Replace the cover strips.

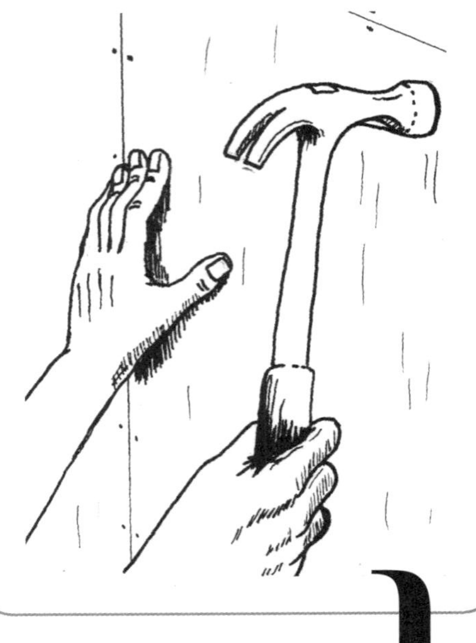

8

Fill the holes with putty and paint them to match the rest of the wall.

Cleaning walls and ceilings

Dirty finger marks, mould, spiderwebs and other dirty spots should be removed from walls and ceilings as soon as they appear to avoid the need for repainting.

What you will need

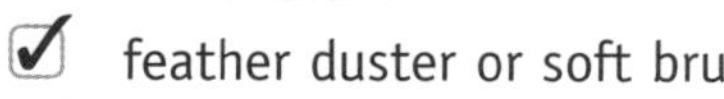

- ✓ feather duster or soft brush
- ✓ scrubbing brush
- ✓ soft cloth
- ✓ abrasive powder
- ✓ cleaning rags or old towels

INSTRUCTIONS

1 Remove any pictures or wall hangings.

2 Use a feather duster or a soft brush to remove spiderwebs and dust. Begin at the top of the wall and work downwards.

3 Use a scrubbing brush and a damp soapy cloth to wash the wall. Begin at the top and work downwards.

4 Use a soft cloth and abrasive powder to remove any dirty spots.

5 Rinse the wall with clean water. Dry off the excess moisture by using cleaning rags or an old towel.

6 When the wall is dry, rehang pictures and wall hangings.

Repainting walls

Painting makes a wall look more attractive. It also protects the surface from wear and tear so that it lasts longer and does not require painting so often. Timber walls can be stained, varnished or treated with chemicals.

What you will need

- abrasive paper
- wood filler or putty
- dust cloth
- paintbrush
- undercoat paint
- glossy or semi-glossy paint

INSTRUCTIONS

1 Using abrasive paper, scrape away any dirty spots and old paint.

2 Use wood filler or putty to fill in any cracks in the wall.

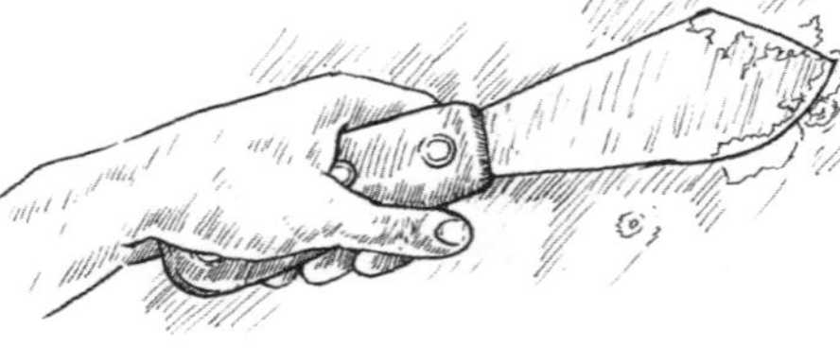

3 Wipe away the dust and dirty spots with a dust cloth.

4 Apply undercoat paint to the wall and let it dry.

5 If you are painting an external wall, apply a top coat of glossy paint. On internal walls you should use semi-glossy paint for the top coat, because paint that is too glossy will reflect light and hurt your eyes.

Replacing blind walls

Blind walls are commonly used in traditional houses in Papua New Guinea. Traditional blinds are woven from pitpit, bamboo and the midrib bark of the sago palm (pangal). These blinds are hand-woven and often have very beautiful and distinctive designs.

Blinds are used on both external and internal walls. Blinds are best used inside where they last longer and their beauty can be appreciated.

If blinds are used on external walls that are exposed to the sun and rain they will lose their colour and eventually break and decay. Rotten and broken blinds must be replaced with new blinds.

What you will need

 cutting tool

 tape measure

 hammer and nails

INSTRUCTIONS

1

Cut loose the ropes that hold the blinds to the wall posts. Pull out the nails and the materials used as cover strips.

Replacing blind walls continued ...

2

Pull the blind away from one end and roll it to the other end.

3

Measure the old blind and cut the new blind to the same size.

4

Place one end of the new blind on the wall post and roll the blind to the other side.

5

Tie the blind firmly onto the wall posts or nail on timber pieces as cover strips.

Replacing a pangal wall

Pangal blind walls should be replaced when they start to rot.

What you will need

- ☑ cutting tool
- ☑ hammer and 2-inch nails

INSTRUCTIONS

1 Cut the ropes loose.

2 Remove the pangal one by one.

3 Tie the new prepared pangal onto the wall posts using a new vein.

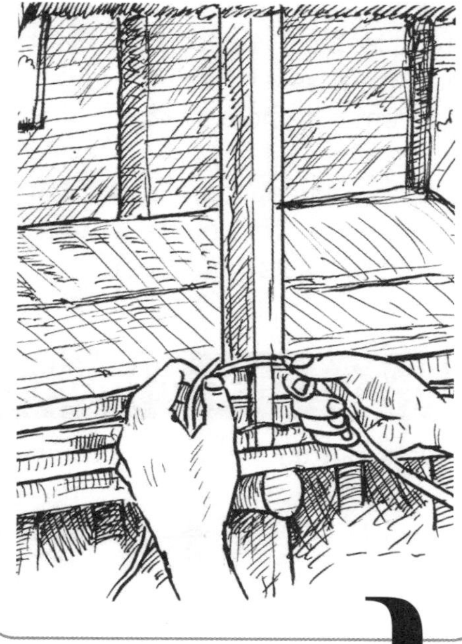

4 You can also nail the pangal onto the wall posts using 2-inch nails.

Maintenance to windows

In the past, many houses were built with no windows. In many areas this was done for security, and in the Highlands it was done for warmth.

Today most people know that it is important to have windows for health reasons. Windows let fresh air and sunshine into the house, which kills germs.

The most common type of window in modern houses is the louvre. Louvres are made from metal and wood.

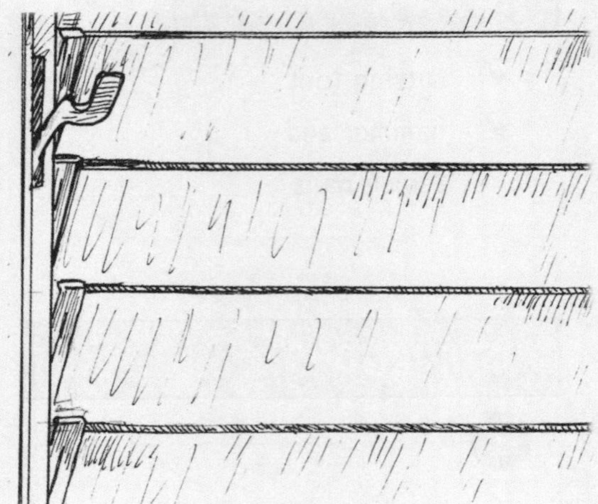

How to clean louvres

Louvres can get very dusty so they should be cleaned regularly.

What you will need

- ✓ feather duster
- ✓ soft cloth
- ✓ abrasive powder

INSTRUCTIONS

1. Remove the curtains or tie them back.

2. Remove loose dust from the louvres with a feather duster or soft cloth. Start at the top and work downwards.

How to clean louvres continued ...

3

Wash each louvre with a cloth and soapy water. Always begin working at the top.

4

Use a soft cloth and abrasive powder to remove any hard dirt.

5

Rinse each louvre by wiping it with a cloth dipped in clean water.

6

Dry and polish glass surfaces with newspaper or a soft dry cloth.

How to replace louvres

If louvres become damaged they should be replaced so that the window can still be opened and closed.

What you will need

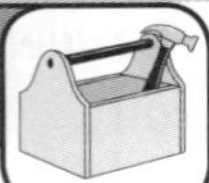

- ☑ tape measure
- ☑ saw or glass cutter
- ☑ screwdriver and screws

INSTRUCTIONS

1 Remove the curtains or tie them back.

2 Open the window so that the louvres are in a vertical position and remove the broken louvres.

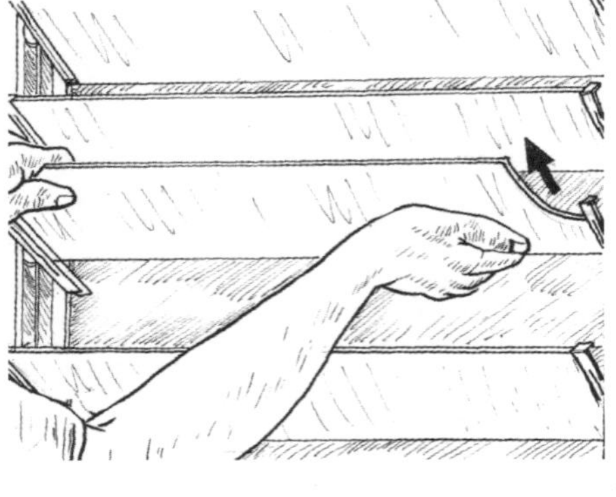

3 After removing the damaged louvres, transfer the measurements to a new piece of glass or wood and cut out a new louvre.

4 Place the new pieces into the louvre frames and screw them in firmly.

Maintenance to floors

Repairing a squeaking floor

One of the most common problems with floors is a squeaking floor.

What you will need

- ✓ hammer
- ✓ nails

INSTRUCTIONS

1

First check if the boards are loosely nailed and re-nail them firmly.

2

If several boards are loose, the joist may be sagging (sinking) due to too much weight. Restore it to its original level with a temporary support. Nail on a piece of timber that is the same size and long enough to hold the joist.

Replacing damaged floorboards

If the floorboards are badly damaged you will need to cut out the worn section and replace them.

What you will need

- ✓ bolster
- ✓ hammer
- ✓ nails

INSTRUCTIONS

1 If the floorboards are square-edged, insert a bolster near the end of one board.

2 Angle the bolster away from the board and tap it down with a hammer.

3 Raise the end of the board up until a claw hammer can be pushed beside the bolster.

4 Lever the hammer and ease the bolster along the board to lift it up.

5 When the first board is lifted, use the claw hammer to lift the others. Lever each board gradually, taking care not to split the wood.

6 Replace with new boards and nail down firmly.

Sweeping, mopping and polishing floors

There are three ways to clean floors:

- sweeping
- mopping
- polishing.

What you will need

- ✓ broom
- ✓ mop or damp cloth
- ✓ disinfectant
- ✓ sandpaper
- ✓ paint, floor polish or varnish

INSTRUCTIONS

1 Sweep the floor thoroughly.

2 Using disinfectant in water, wash the floor with a mop or a damp cloth.

3 Rinse and wipe the floor clean and let it dry.

4 To polish floors, sandpaper thoroughly and wipe the floor surface clean from dirt and dust.

5 Apply paint, floor polish or varnish to the floor.

Repairing bamboo floors and sago palm stem floors

Floors in traditional Papua New Guinean houses are made from bamboo, strong pitpit and sago palm stems. Blind floors are used in traditional Highlands houses that are built on the ground.

What you will need

 hammer

- ✓ nails

INSTRUCTIONS

1 Remove the damaged section of the floor.

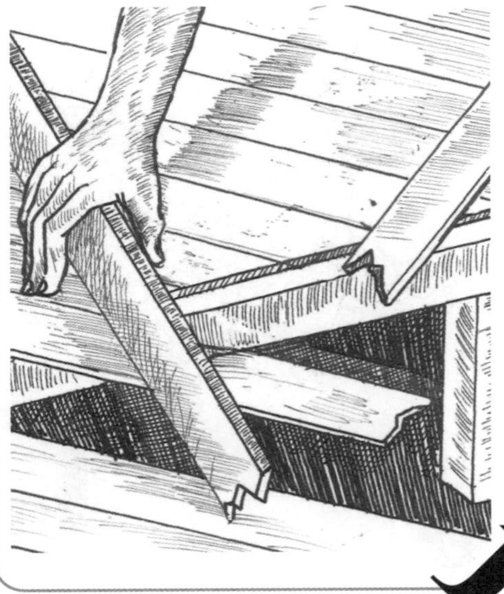

2 Replace the section with a new section of the material used for the whole floor.

3 For a blind floor, remove the whole blind and replace it with a new floor blind.

Other common home maintenance

Cleaning the bathroom and toilet

Dirty toilets contain germs that can make you sick. It is important to clean your bathroom and toilet regularly so that you stay healthy.

What you will need

- disinfectant
- damp cloth
- toilet brush
- cleaning powder or paste

INSTRUCTIONS

1. Using disinfectant in water, wipe the surfaces of the shower, hand basin and toilet with a damp cloth. Disinfect the cloth after wiping the toilet.

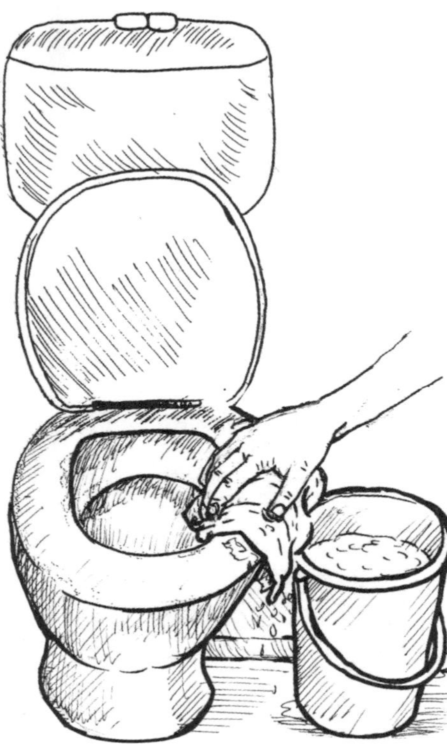

Cleaning the bathroom and toilet continued ...

Apply a cleaning powder or paste to mould or hard dirt.

3

Use a toilet brush to scrub inside the toilet bowl.

4

Rinse the surfaces with water and wipe them clean.

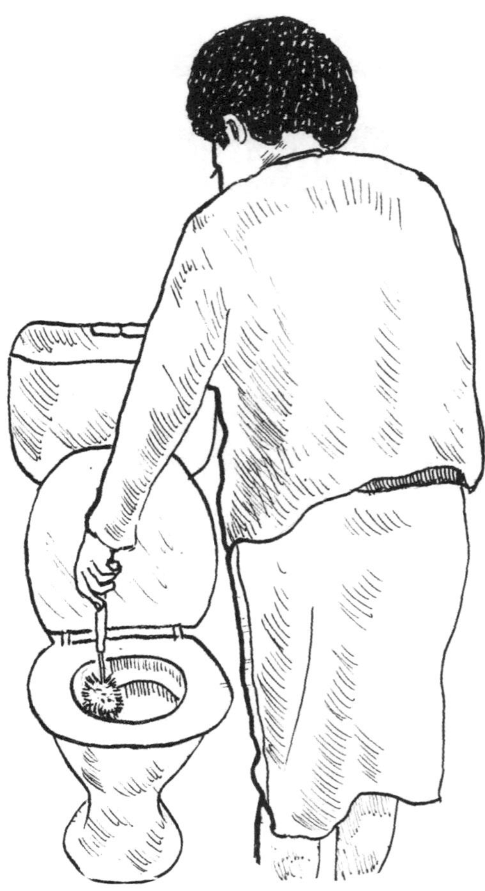

Repairing door hinges

Frequent opening and closing of doors can cause door hinges to come loose.

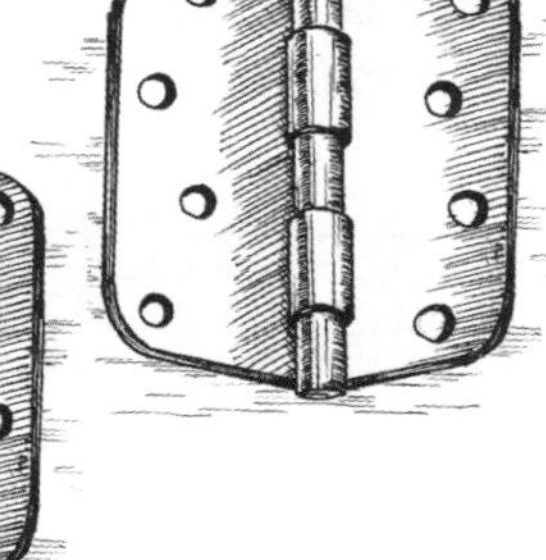

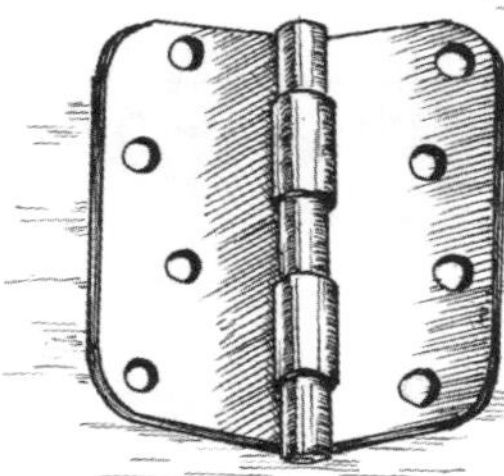

What you will need

- ✓ screwdriver
- ✓ oil

INSTRUCTIONS

1

Tighten the hinge screws firmly.

2

Apply oil onto the parts of the hinge that cause friction.

3

If this does not fix the problem, unscrew the door hinges and replace them with new hinges.

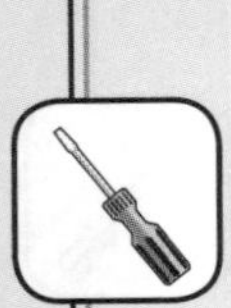

Repairing door locks

Many doors have deadlocks. These sometimes come loose and need to be repaired.

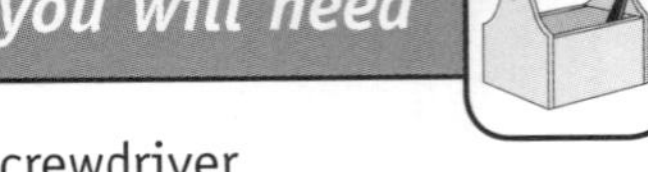

What you will need

- ✓ screwdriver

INSTRUCTIONS

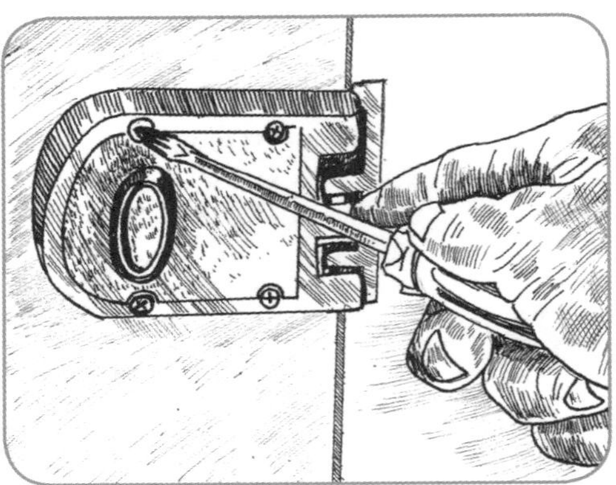

When the deadlock comes loose, firmly tighten the screws.

***Note:** If there is a problem inside the lock, see your local locksmith.*

Repairing doorsteps

The steps leading up to a building are an important part of all buildings. People use these steps every time they go in and out of the building.

Steps must be checked regularly to ensure that the joins are not loose and the steps are in good order.

What you will need

 hammer

 spanner

INSTRUCTIONS

If you notice that the steps are coming loose, reinforce them by nailing the joints and tightening the bolts and nuts.

Building a new doorstep

If the steps are broken or worn out you will need to replace them with new steps.

What you will need

- ✓ tape measure
- ✓ saw
- ✓ hammer
- ✓ nails
- ✓ paintbrush
- ✓ paint

INSTRUCTIONS

1. Remove the steps of the building.

2. Dismantle the stepping timber from the two side rails.

3. On removing the timber, take its measurements and transfer these to new pieces of timber. Cut the timber to the right size.

4. Position the new pieces and nail them together firmly.

5. If the step is a short step, nail the pieces together, move the completed step to the building and join it to the building.

Building a new doorstep continued ...

6

If the step is a long step, first join the two side rails firmly to the building. Then attach the other pieces to the side rails.

7

Apply paint to the steps to prevent timber from rotting quickly.

Furniture maintenance

Some of the types of furniture we like to have in our houses or schools are:

- beds to sleep on
- chairs to sit on
- tables to write and eat on
- cupboards to store goods in
- shelves to store books on.

It is important to look after the furniture you have. Furniture is expensive, so it is cheaper to repair broken and damaged furniture than to buy new furniture.

Here are some general points to consider when planning maintenance to furniture:

- Carefully study the damaged item to see what materials it is made of. Most household furniture is made of wood, plastic or metal.
- Take note of the type of joints used in making the item.
- Think of the tools used to make the item and the tools you will use to mend the item.
- Assess the state of the damage and think of the materials you will need to mend it.
- Plan how to carry out the maintenance.

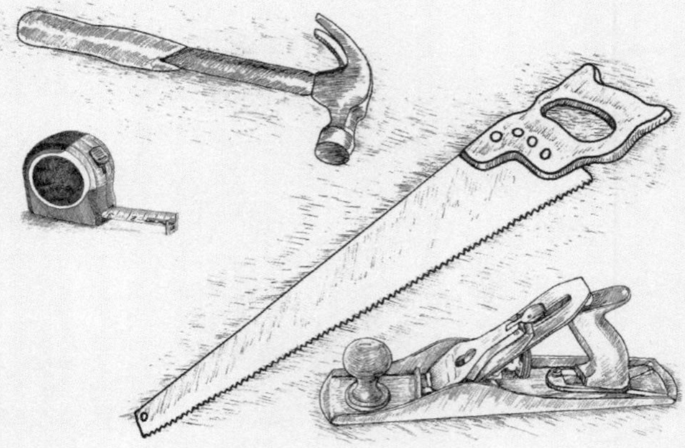

How to repair loose joints

Sometimes the joints of a piece of furniture will come loose over time, especially if it is a piece of furniture you use regularly.

What you will need

- ☑ glue (PVA or contact adhesive)

- ☑ screwdriver and screws
- ☑ hammer and nails

INSTRUCTIONS

1

To fix loose joints, apply glue to the joints. Use PVA glue for wood and contact adhesive glue for plastic.

2

Clamp the joints together tightly until the glue is dry.

3

Screw or nail the joints together.

How to repair broken furniture

If the furniture is broken, you will need to replace the broken piece with a new piece of the same material.

What you will need

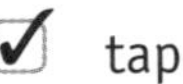

- tape measure
- saw
- hammer and nails
- sandpaper
- screwdriver and screws
- paintbrush and paint

INSTRUCTIONS

1 Remove the broken piece from the piece of furniture.

2 Measure the broken piece and cut out a new piece of the same material.

3 Put the new piece in place and nail or screw it on firmly.

4 Sandpaper the whole piece of furniture and wipe it clean.

How to repair broken furniture continued ...

5

If the legs are metal and are rusty, sandpaper and brush them thoroughly to remove the rust.

6

Paint the whole piece of furniture.

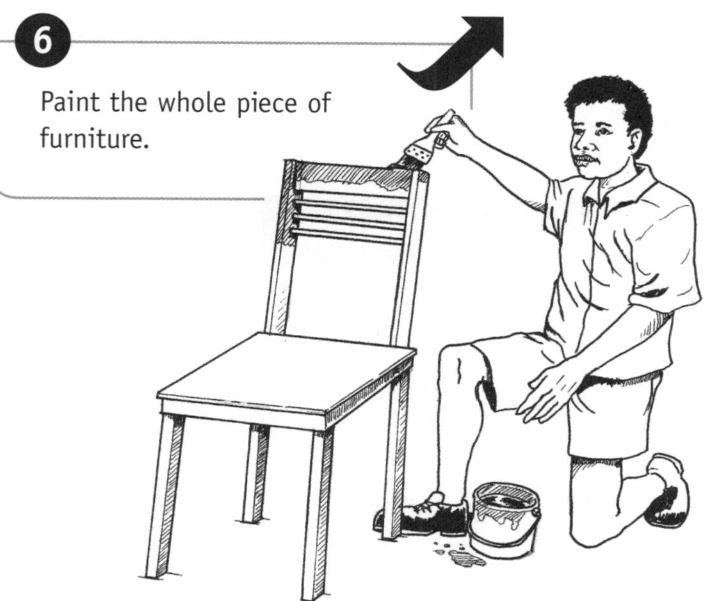

7

Leave it to dry in a place free from direct sunlight and dust.

Plumbing maintenance

Plumbing in homes and schools can cause major problems if simple faults are not repaired. If a simple fault is not attended to, it could become a bigger problem that will require major and expensive maintenance. The problem could also reduce water supply and become a health hazard if left untreated. Therefore it is extremely important to repair plumbing faults promptly.

For major plumbing maintenance, call a plumber to carry out the repairs. Where a plumber is not available, a good grasp of how your plumbing system works will help you to carry out your own repairs.

Repairing leaking water taps

Leaking taps can be fixed by replacing the tap washer.

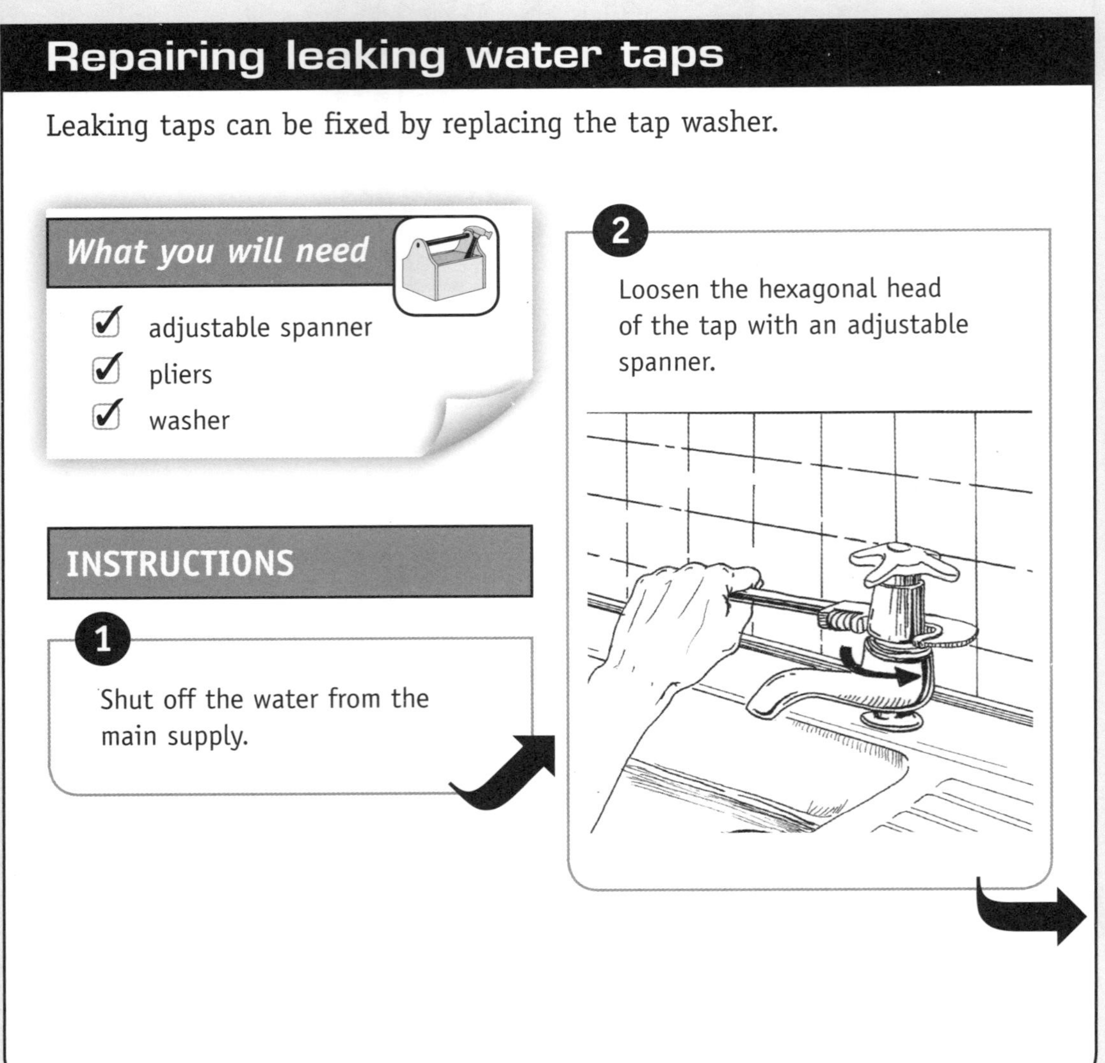

Repairing leaking water taps continued ...

3

Unscrew and loosen the nut by hand.

4

Lift off the top of the tap.

5

Using pliers, grip the edge of the jumper valve. On most taps the jumper valve fits loosely and will just pull out.

6

Undo the nut and take off the old washer. Check the valve seat inside the tap.

7

Fit the new washer and replace the metal washer and nut. If the thread is damaged, get a complete new valve.

8

Reassemble (put back together) the other parts.

Note: *If the washers need frequent replacement, the valve seat may be damaged. If this is the case, remove the whole tap and replace with a new tap.*
This procedure is also used to replace leaking taps in sinks, showers and toilets.

Repairing blocked sinks

The most common plumbing problem experienced by households is a blocked sink. When a serious blockage occurs it is wise to call a plumber. If a plumber is not available, follow these steps.

What you will need

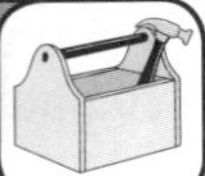

- ✓ plunger
- ✓ wire
- ✓ drain auger
- ✓ bucket

INSTRUCTIONS

1 Remove the sink stopper.

2 Use a plunger on the drain hole to try and remove the blockage.

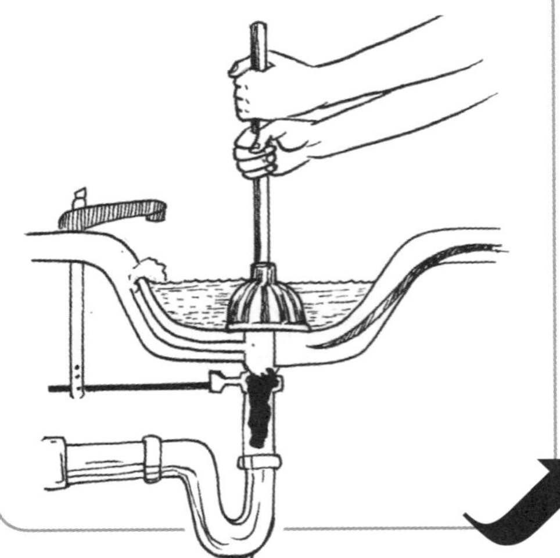

3 If this doesn't work, remove the clean out plug from the pipe under the sink and use wire to fish out the blockage.

Repairing blocked sinks continued ...

4

If the sink is still blocked, remove the curved section of pipe.

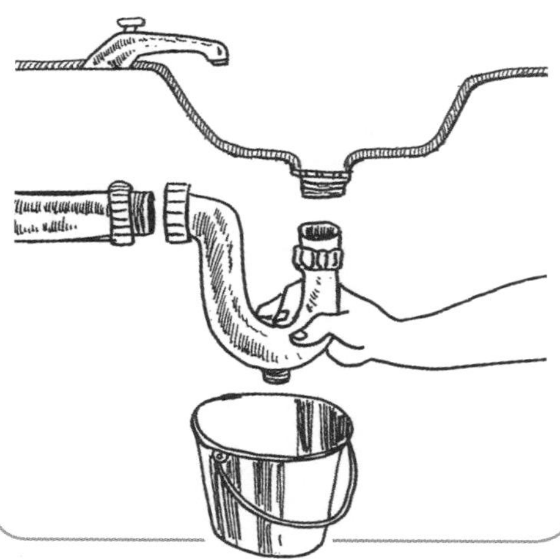

5

Insert a drain auger in the drain line. Rotate it and work it in until the blockage is removed.

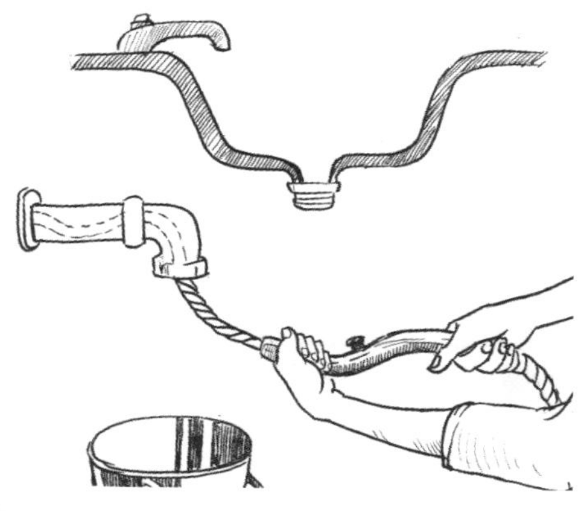

Repairing blocked toilets

When you flush the toilet and the water level in the bowl rises rather than recedes, follow these steps.

What you will need

- ✓ plunger
- ✓ drain auger

INSTRUCTIONS

1

Place the plunger over the discharge opening in the toilet bowl and work it up and down.

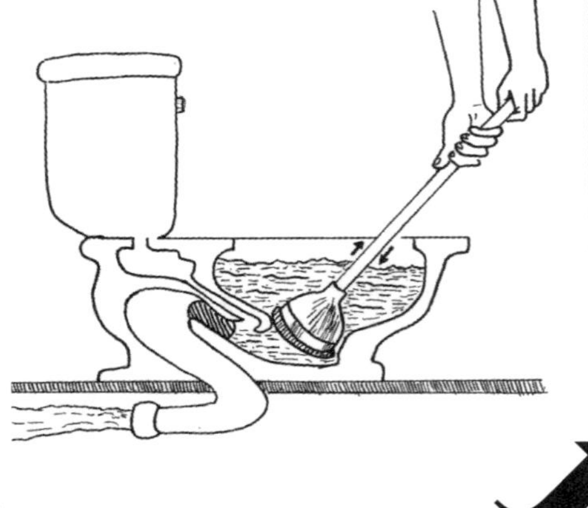

2

If Step 1 does not work, use the drain auger and work the snake into the drain. Crank it in one direction until it becomes tight, then pull it back. Often it will bring the obstruction with it.

3

If all else fails, remove the toilet from the floor, turn it upside down and work the obstruction out through the discharge opening.

Repairing broken water pipes

A broken water pipe can waste a lot of water, so it should be repaired as soon as possible.

What you will need

- ✓ hacksaw
- ✓ tape measure
- ✓ pipe glue

INSTRUCTIONS

1. Turn off the water at the main tap.
2. Cut out the broken section of the pipe with a hacksaw.
3. Measure the broken piece of pipe and cut a new piece of pipe the same size.
4. Apply pipe glue to both ends of the new pipe.
5. Fit the new section around the outside of the top pipe and the inside of the bottom pipe.
6. Leave the pipe joins to dry completely.
7. Once the glue is dry, turn on the water to see if there is any leaking from the joins.
8. If water leaks from the joins, wrap plumber sticking tape around the joins.

Safety and first aid

It is important that anyone doing maintenance and repairs is familiar with basic first aid procedures, as accidents may happen. Every person should also have a basic first aid kit. The most common injuries to maintenance workers are caused by using tools carelessly or inappropriately.

These injuries can include:

- wounds
- excessive bleeding from wounds
- eye injuries
- broken bones.

Some basic first aid procedures

Wounds

A wound is a break in the skin caused by an injury. Cuts, grazes and burns are all wounds. Wounds are dangerous because they can cause bleeding and infection. Infection can be prevented by proper treatment of wounds.

What to do

- Stop any bleeding quickly (see the section on bleeding for more information).
- Small cuts and grazes may be cleaned with soap and water or an antiseptic solution before a clean dressing such as a clean towel, sheet or laplap is applied.

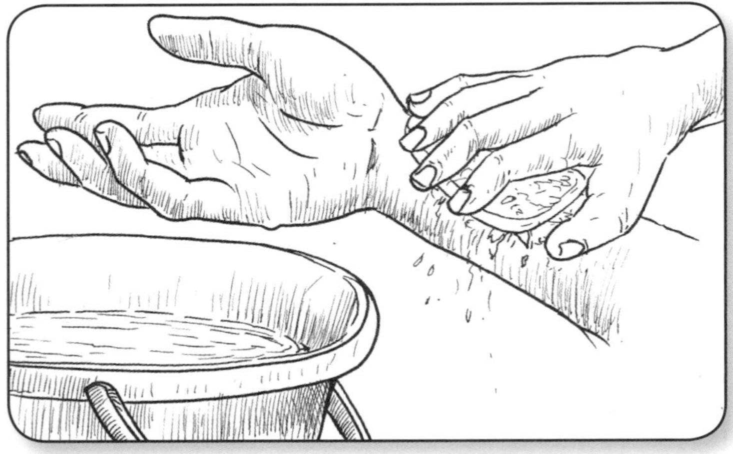

- Do not clean a wound that is bleeding heavily. Stop the bleeding if you can and take the victim to a health facility.

- Use plastic bags on your hands to stop you coming into contact with any blood. Dispose of plastic and any soiled dressings into a sealed container, and either burn it or bury it well.

- A puncture wound, such as an injury from a rusty nail, may become infected later. It is wise to see a health worker, because an anti-tetanus injection may be required.

Bleeding

If too much blood is lost from the body through a wound or injury, the victim may become seriously ill and even die. Always stop the bleeding as soon as possible.

What to do

- Stop the flow of blood by pressing the skin edges together and placing a bulky dressing, such as a clean towel, sheet or laplap, firmly on the wound.

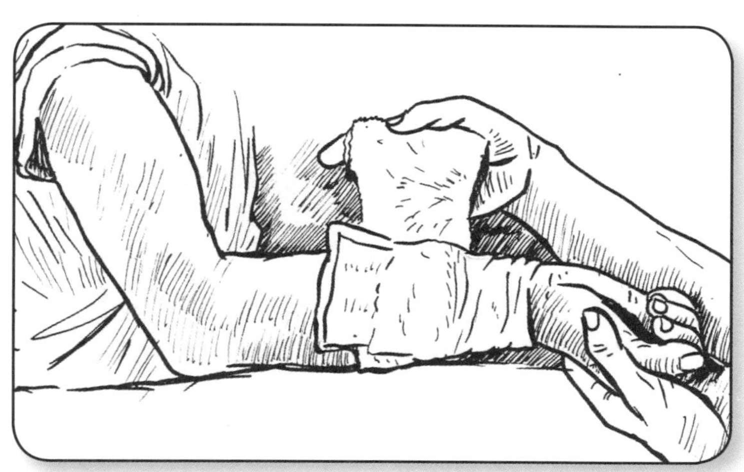

- Keep firm pressure on the wound for ten minutes, with the victim in a resting position. If the wound is on a limb, the limb should be raised to slow down the blood flow.

- If bleeding continues through the first dressing, put another dressing over the first dressing. Do *not* remove the first one.

- Keep the injured part raised for a while after the bleeding has stopped. A health worker should check severe wounds.

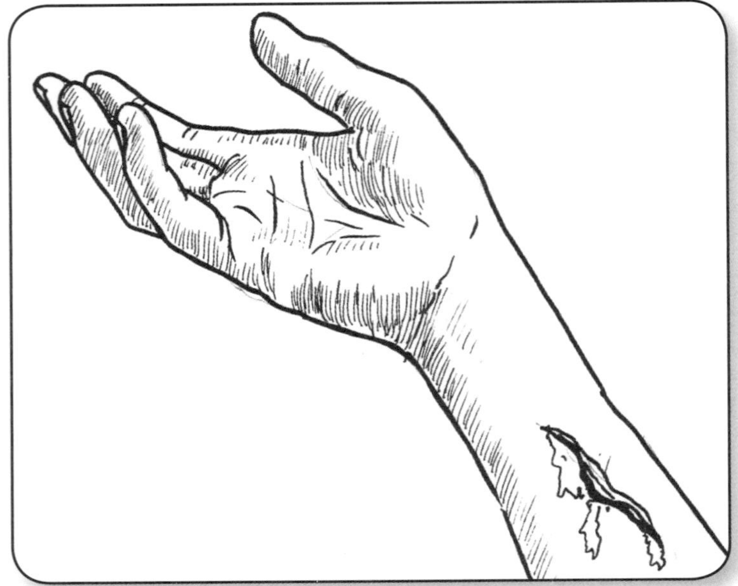

- Foreign bodies such as glass, metal or splinters can be removed if they can be seen clearly on the surface of the wound. Don't try to remove anything that is deep in the wound.

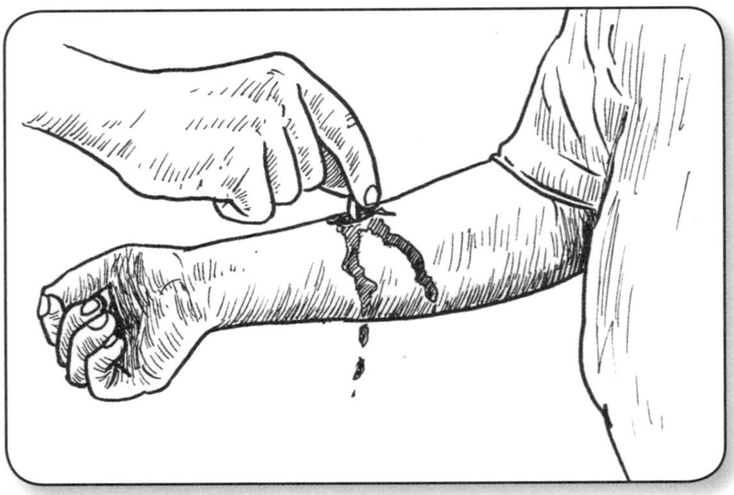

- Never use dirt, kerosene, lime or coffee to stop bleeding.

Eye injuries

Great care is needed when a foreign body gets into the eye. Rough treatment may lead to permanent loss of vision.

What to do

- Help the victim flush the eye under cool, running water. This may dislodge the foreign body.

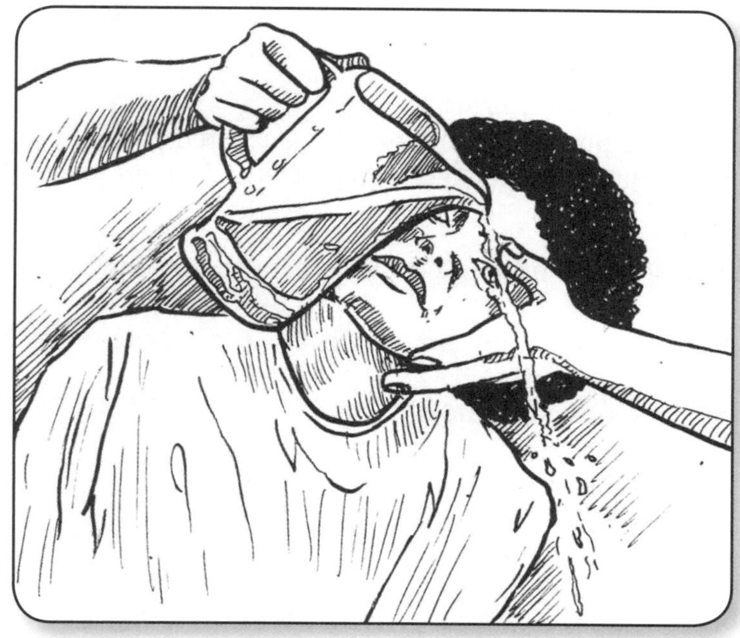

- If this is unsuccessful, gently pull down the lower eyelid and look for the object.
- If it can be seen on the eyelid, or on the white part of the eye, gently remove it with a corner of a clean handkerchief or tissue. *Never* use sharp objects to remove the foreign body.

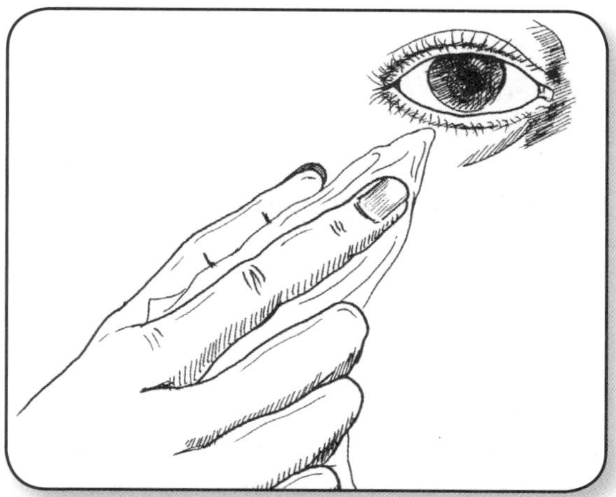

- If the object cannot be seen or is present over the coloured part of the eye, *do not try to remove it*. Cover the eye with a clean pad and seek medical aid.

Broken bones (fractures)

A cracked or broken bone is called a fracture. It is often hard to tell whether or not a bone is broken. When there is doubt, the injury must be treated as a fracture.

How to recognise a fracture

- pain at, or near, the injury
- swelling, with bruising later
- loss of power or function of the injured part
- deformity or change in shape of the injured part.

What to do

- Handle the injured part with care to prevent further pain and shock.
- Quickly cover any wound with a sterile or clean dressing.
- Support the injured area with padding and bandages as required. Use other parts of the body for support; for example, secure the arm across the chest to support broken ribs.

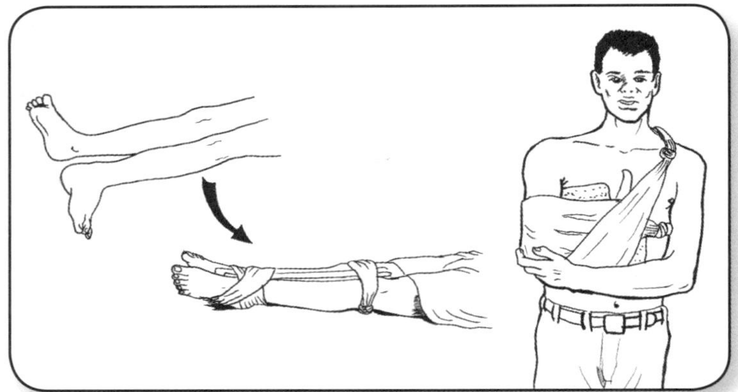

- When the fracture is in a long bone (for example, the shin or the forearm), stop the joints above and below the injury from moving.

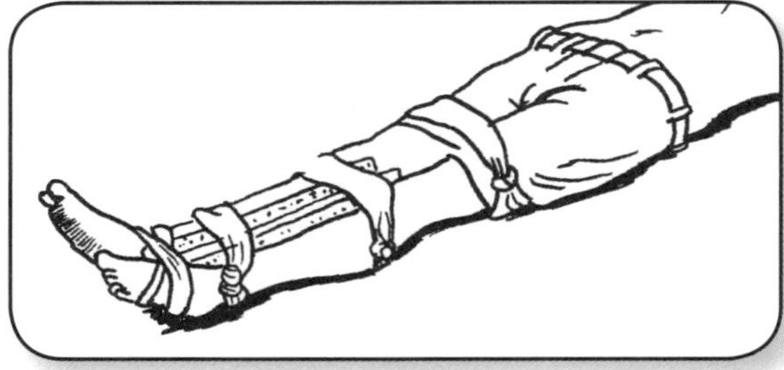

- Additional bandages may be placed above or below the injury but these should never be placed over the fracture.
- Seek medical aid as soon as possible because special treatment may be required.

Glossary

abrasive powder	powder that is mixed in with water and used to scrape dirt off surfaces
bituminous roofing compound (tar)	thick black liquid used for covering roads and roofs
cover strip	piece of timber used to cover the edges of a wall
countersink	enlarge the top of a hole so that head of a screw or bolt fits neatly into it
corrugated iron	sheet iron used for roofs
disinfectant	poisonous liquid added to water and used for cleaning walls and floors
downpipes	iron pipe that is attached to the gutter to carry rain water down to the tanks
furniture	household items such as tables, chairs, stools, desks, cupboards and shelves
gutter	a square-shaped iron sheet nailed to the end of the roofing iron to catch rainwater
kunai grass	a grass that is commonly used as a roof covering
locksmith	person who fixes locks
masonite sheet	flat sheet of board used for walls
morota	woven or sewn sago palm leaves used for roof covering
nogging	short pieces of timber nailed in between the studs to hold the wall firmly
overlap	to partly cover something
pangal	wall panelling made from the midrib bark of the sago palm
plywood	flat sheet of wood used for walls
ridge cap	the top cover of the rooftop
stock book	record book for tools
studs	long pieces of wood used as posts in the wall frames of houses
turpentine	poisonous liquid used to clean paintbrushes
undercoat	the first coat of paint applied to a wall, usually a dull white paint that dries quickly
undersize	smaller than the correct size
valve	device that controls the flow of water in taps

References

Department of Education 1996, *Practical Skills Maintenance Teachers Guide* Grades 7 & 8. Department of Education, Waigani, Papua New Guinea.

Department of Education 2005, *Making a Living Student Books 1 & 2*. Oxford University Press, Melbourne, Australia.

Notes

Notes continued ...

Notes continued ...

Notes continued ...